Kenneth
5/18/98

Biotechnology: The Future of Motherhood and Sex

Kenneth Quade

VANTAGE PRESS
New York

FIRST EDITION

Published by Vantage Press, Inc.
516 West 34th Street, New York, New York 10001

Manufactured in the United States of America
ISBN: 0-533-10652-4

Library of Congress Catalog Card No.: 93-93853
0 9 8 7 6 5 4 3 2 1

Contents

Author's Note

This book is for those who will be living in the next century, the twenty-first century. By the middle of the next century, people living then will know whether I was wise or a fool. But let me remind you: history has shown reality often begins as an absurdity.

Preface

Knowledge is useless when the masses fail to heed but instead ignore that knowledge.

In the mid-1920s, when Gen. Billy Mitchell was a low-ranking army officer, he warned that America would go to war against Japan. All this did was help get him court-martialed. Those sitting in judgment at the time would not believe they, their sons, or their grandsons would be fighting in just such a war in less than twenty years. But "absurdity" became a reality.

In the mid-1950s, a geologist warned that America would be running short of oil in twenty years. It was hard to convince anyone paying one dollar for five gallons of gasoline then that he would be paying one dollar for only one gallon of gasoline in twenty years. But another absurdity became a reality.

Were both men prophets? I say no. They took the facts and evidence already obvious, used common sense, and came to the correct conclusion.

A government report on energy ("Resources for Freedom," also known as the Paley Report) released in June of 1952 recommended the following initiatives: Begin to produce energy-efficient and smaller automobiles. Use renewable resources such as solar energy. Save our domestic reserves of oil and import more foreign oil.

Few people know this report exists. *Time* magazine at the time it was released didn't use it as a cover story. Instead *Time* gave it less than a full page of coverage. Twenty years later, magazines, including *Time,* frequently published cover stories on the issue of energy.

After giving testimony before a congressional committee on the trans-Alaska pipeline in the spring of 1973, I was asked by the chairman of that committee to submit a brief critical analysis of the final environmental impact statement on the pipeline. I said I thought much of the statement was nothing more than a brainwash right out of the public relations department of the oil industry. I was especially disturbed and agitated by the claim of the oil companies that they would deliver North Slope crude oil to California for $3.29 to $3.33 per barrel.

With oil wells drilled in the Lower Forty-eight costing only a fraction of those wells drilled on the North Slope, it was obvious North Slope oil would increase crude oil prices and thus retail prices to consumers. Add to that high transportation costs, with Alaska's climatic conditions. When the oil companies stated the pipeline would cost $6 billion to construct, I publicly challenged that statement with letters to newspapers and said it would cost $10 billion. The final cost came close to my estimate.

The oil companies also made the outrageous claim that transportation costs through the pipeline would only be one dollar per barrel. When I called this to the attention of the media and said the cost of North Slope oil, including drilling and transportation, could increase prices to around ten dollars per barrel, I was met with laughter. A tripling of current crude oil prices? How ridiculous! The jump, only months later, to $11.50 per barrel during the Arab oil embargo came as no surprise to me, even though it did to the rest of the country.

The government report released in 1952 ("Resources for Freedom"), which I read more than once, stated that the domestic cost of crude oil production in the United States would determine world prices. Why should other oil-producing countries sell their oil at a cheaper price than that of what the largest oil-consuming country in the world, the United States, could produce domestically?

The oil companies' calling the Arabs villains at the time of the embargo was all an act, similar to President Roosevelt acting surprised at the attack on Pearl Harbor by the Japanese. The trans-Alaska pipeline never would have been built without a large increase in gasoline and fuel oil prices. The oil companies knew that right from the beginning.

The politicians in Washington and the American media's failure to challenge the oil companies before that became a reality was the result of either their ignorance or their suppression of the information—just another example of manipulating the masses. Is the individual meant to be only a pawn in an economic system from womb to tomb?

As for the media, when I am asked my opinion on an issue I often respond by saying, "I'm ignorant. I only know what the media are telling me."

As for the politician, it was former senator Howard Baker, appearing as a guest on a Sunday morning news panel program, who stated it concisely and well. He told the journalists on the panel asking the questions, "You know, you journalists and we politicians are cut from the same side of the cloth."

It was during my involvement in the trans-Alaska pipeline controversy that I also became involved in genetic engineering and biotechnology. In the fall of 1973 I was a guest on a radio talk show in Anchorage, Alaska. I told the listeners we could expect at any time a woman that had no sexual contact with a man would give birth to a child by the means of a procedure called in-vitro fertilization, and it would probably happen first in England. As a result, I said, some women would become substitute mothers for a fee of $10,000 to $20,000.

I remember vividly when a listener called in and told the host he had a real crackpot for a guest. Within five years what the media called a test-tube baby was a reality. Not long after, what the media labeled as surrogate motherhood (which I had referred to earlier as substitute motherhood) became a reality, and at a fee I correctly predicted.

I've read very little science fiction. Nor do I consider myself a prophet or a futurist. Having read enough government and private reports on the subject of genetic engineering, I had come to an accurate conclusion. With the people depending on the media for their information, naturally they were the last to know.

Maybe the public wants to be the last to know. Ignorance frees one from responsibility. Wasn't that the defense the German people used after the war regarding the ovens?

I've used the two examples of Billy Mitchell and the geologist, plus my two experiences, hoping to give some credibility to what is to follow. More people are beginning to realize in this age of high-tech and ever-changing life-styles that the word "never" should only be used to say "never say 'never.' "

I'll be asking uncomfortable and unsettling questions in the following pages. The questions will be unpopular, just as profound and thought-provoking questions were unpopular during the age of Socrates.

As Bertrand Russell wrote in his time: "The central problem of our age is how to act decisively in the absence of certainty." That seems to apply to all ages.

Although I believe discussion and debate of all and not just a few of the implications of biotechnology should already be under way, I'm probably in the midst of a society similar to the one trying to convince people five hundred years ago that the earth was round. One listener on a radio talk show accused me of comparing apples and oranges with that example. I assure you, biotechnology is going to make fruit salad out of those apples and oranges.

I firmly believe much that will come about resulting from breakthroughs in biotechnology will be caused by one factor:

economics! Or as some would say, the three most important issues in a presidential election are: the economy, the economy, and the economy.

American society has become so accustomed to a high standard of living and trying desperately to achieve that standard of living that some of our most cherished and sacred institutions and traditions will be discarded. In order to compete globally, whether it be economically or just due to our desire to be number one, some of it will be forced upon us.

Self-preservation will play a part in sacred institutions being eliminated. Are overpopulation and other unknowns out there going to determine what direction civilization will or must take? A direction that few living today can comprehend, the same as few, if any, one hundred years ago could have comprehended the way we live now.

Most, if not all, will be offended, even angered, by what they are about to read. It will lose me friends and alienate strangers. So be it.

Biotechnology and the Future of Motherhood

Does society have the right to deny children healthier bodies when better ways to produce the human species are found, including an artificial womb that will render pregnancy obsolete?

This is a question some people living today will have to come to grips with in their lifetime. Circumstances will prevail where this can no longer be avoided.

Some president will have to make the decision whether American society will be a leader in perfecting and utilizing an artificial womb, just as an earlier president challenged us to be the first to land a man on the moon. Another *Apollo* is waiting in the wings, this time biological. The race is on.

Fetal research toward producing a human embryo and extending its life outside the body is the front end and beginning of an artificial womb.

As for the other end, we can now save premature babies born after only half the normal gestation period has passed. When the two meet, that's when obsolesence of the human female uterus begins.

Why, you ask, would anyone, any nation, want to experiment in such an area?

Challenge! Ask mountain climbers why they climb a mountain, and some will answer, "Because it's there." The scientific community is no different. As a college student studying science said, "I just like the idea of fixing people and making them better."

The direction American society is heading in encourages that challenge. Thousands upon thousands of defective children are born each year because women contaminate their wombs with drugs, venereal disease, alcohol, and smoke. One has to cringe at seeing a newborn baby in a hospital nursery going through drug withdrawal because of an irresponsible mother. Is it moral and ethical to ignore what present society is doing to future generations?

In the end, though, it will not be morals and ethics that determine action. It will be that one word again: *economics*! As former president Harry Truman said, "The most sensitive nerve in the human body is the pocketbook nerve."

The pocketbook nerve is becoming more sensitive every day in the American people. They are becoming concerned that the American Dream is dead: that their children will never enjoy the standard of living their parents experienced.

It's been reported that more than 350,000 babies each year are born addicted to drugs. It has also been stated that it costs society more than $200,000 to rehabilitate from birth to kindergarten each child born addicted. It costs even more to care for an AIDS baby from birth to death.

No matter how you add it up, either with pencil and paper or with a calculator, 350,000 times $200,000 comes to $70 billion. Once the masses realize what an immense amount this is, in a country concerned about its national debt, balancing the budget, and the threat of a declining standard of living, drastic measures will be taken.

We must begin to wonder if we are alienating a large segment of the baby population from their mothers. What happens when these children discover that the way they are resulted from their mothers' contaminating their wombs? Are we beginning to rear a segment of the present generation that will embrace a new way to produce babies because of the poor beginning they had to endure?

Take Fetal Alcohol Syndrome, referred to as FAS. Here a child is not only born mentally retarded, but can also end up with ugly facial features. How will a sixteen-year-old girl looking in a mirror react to learning her ugliness is the result of her mother contaminating her womb with alcohol? Will the hatred flare when the teenager accuses her mother of "booze-drinking" irresponsibility? Often mother/daughter relationships are already strained without adding another volatile issue.

I remember listening to a talk show on the car radio while driving through Nevada one year. A listener called in who resented her parents for burdening her with bad genes. Add birth defects to this, and lawyers will have a field day with children suing parents.

FAS is adding billions more in costs to an already alarming national financial burden.

Then there is AIDS. These children usually don't live long enough to resent their mothers. But can you blame all mothers in this case? Some acquire AIDS through no fault of their own. Still, much and eventually probably all of the costs for caring for these AIDS babies will be shouldered by the taxpayers.

There are some who say it is nobody's business what they do in the privacy of their bedrooms. As AIDS creates panic, people will begin to challenge that opinion.

Unless they contribute their own blood as a precaution before surgical operation, many live in horror that a transfusion of blood from a stranger might be necessary, because of AIDS. Many good and responsible people in the health care professions are giving up careers because of AIDS. How many innocent people might have their reputations destroyed because they died of an undisclosed illness and unfounded rumors claimed it was AIDS? Even some barbers won't shave your neck after a haircut because of AIDS.

I've only listed a few examples of what this fear is doing to our society. There are many more, but I've made my point.

As this fear leads to panic and hysteria, and more panic and hysteria, measures inconceivable now could be taken later in order to produce babies in artificial wombs and curb the sex drive.

When hijacking of airplanes became a crisis, the solution was to force all passengers to go through a security check before boarding a commercial aircraft. If a small minority—but larger than any estimate of potential hijackers—of women insist on failing to use responsibility in producing babies, then what might happen? Especially if their life-style means an enormous tax is heaped upon middle-class Americans.

Certainly if society continues on its present course, eventually strict restrictions will be forced upon women to prevent social chaos. Genetic screening to weed out bad genes that could result in genetically defective children will be one course of action. Another would be a very regimented and restrictive prenatal care program to prevent birth defects.

Would women want such a program of childbearing forced upon them, and probably more by peer pressure than by government decree? In a diverse society such as ours, could such a program be implemented, regardless of how necessary it was? Would an artificial womb then be more readily accepted?

The artificial womb could make the female womb obsolete by the middle of the next century. By then society may frown on women trying to bear children the old-fashioned way. They would be a minority, just as the Amish are today.

Flying on a commercial airliner more than seven miles above the earth and at a speed of over five hundred miles an hour, I sat next to another passenger who had on his lap a miniature covered wagon he had purchased at some gift shop. Imagine being a member of a wagon train one hundred years ago. Suppose, too, while sitting around the group camp fire one evening you told others that their grandchildren and even in some cases their children would experience just such a thing as flight. Why, they would have put you on a horse, given you some flour and water, and sent you on your way, diagnosed as crazy.

Later, after I changed planes, I made a telephone call from that plane to earth.

The female womb will never be able to compete with such sophisticated technology, and especially what's coming in biotechnology. As we moved out of the age of the horse and buggy into the age of the automobile and air transportation, so are we about to leave one age of human reproduction and enter another. In a way we are now pioneers in the initial stages of this mind-boggling area.

Does this mean we will turn to a form of civil sterilization? When you add up all the various forms of contraceptives, including vasectomies for males and tubal ligation for females, it appears people are already on that course voluntarily.

A young woman, a member of a television audience, said maybe sterilization is necessary in order to save the planet. How many people feel that way?

What some feel may be necessary from an ecological standpoint others may feel is necessary for them to acquire a higher standard of living. It is my observation that while some may take environmental concerns seriously, others would sooner die than give up the prosperity and affluence to which they have become accustomed.

Will we see more and more morals and ethics playing only a supporting cast to what society has concluded is most important: money? Again let me repeat the three most important issues that determine the outcome of elections: the economy, the economy, and the economy.

We must remember, too, that women's wombs are being contaminated with things beyond their control, such things as chemical pollution and radiation for example. These are even affecting male sperm.

Will this eventually lead to the necessity to produce life another way other than from the female womb? With increased space travel and exploration of the universe, can the female womb produce the kind of individual needed for that kind of future? Can the family unit?

When I'm a guest on a radio talk show, usually some listener will call in asking whether the purpose of developing an artificial womb is the same as Hitler's goal of creating a master race. His "love farms" as an attempt to achieve a master race and our present sophisticated eugenic advances of in-vitro fertilization, embryo transplants, and surrogate motherhood are as different from each other as a Model-T Ford is from a modern-day sports car. If he would have had those procedures available then, we would be living with the results today. Think of the consequences of his using eggs from superior German women and sperm from officers of his SS Corp and using women, German or those of conquered countries, as surrogates on a massive scale to carry the resulting embryos to term, as can be done presently.

Perfecting an artificial womb won't take the economic resources that were needed to go to the moon. Any small populated country such as Australia—already a leader in the area of test-tube babies and frozen embryos—could achieve the mind-boggling breakthrough of perfecting an artificial womb. Some country will do it!

A 1972 report to a congressional Subcommittee on Science, Research and Development asks this thought-provoking question about a master race: "Sooner or later one human society or another will launch out on this adventure, whether the rest of mankind approves or not. If this happens, and a superior race emerges with greater intelligence and longer lives, how will these people look upon those who are lagging behind? One thing is certain: they, not we, will be the heirs to the future, and they will assume control."

Are we ignoring this at our peril? Or are we embarking on a new era where the scientific/medical/industrial complex is replacing the military-industrial complex? Is this what is needed to keep us up there as a world leader?

Could the height of irony be that Germans, who were once condemned for trying to achieve a master race, find the term *improvement of the species* more acceptable in the modern age?

If restrictions are put on Germany's military aggressiveness, will this aggression be put into biotechnology? Is this what it will take to maintain an economic superiority and control an increasing population in a small landmass?

Expansion was one of Hitler's goals: get the German people more land to obtain needed resources. Today German prosperity is flooding Germany with immigrants and still the landmass remains as small as before the war.

If you restrict a nation militarily, can you restrict that nation from expanding into biotechnology such as in-vitro fertilization, frozen embryos, and eventually an artificial womb if that nation wants to improve biologically in order to solve its problems?

Then there is Japan. It has already stated that it plans on being a leader in biotechnology. With its desire it seems to control the world economically. Could biotechnological superiority be achieved this way? And certainly Japan is not growing any larger in terms of landmass.

If increased population should begin to lower Japan's

wealth and head it in a third-world direction, would Japan turn to an artificial womb as an answer to this kind of crisis?

Another irony: The two losers in World War II may take the world down a path that those who fought and died in that war never dreamed would happen. If there are Toyotas in garages now, will there be babies from incubators made in Japan later?

As for an artificial womb stabilizing and reducing overpopulation, what about China? It is said one of the reasons it finally turned to the West was to gain some of our scientific know-how, including genetic engineering.

China's plan of one child per family is not working. Will they think an artificial womb is the answer?

If China's course of action is not determined by a belief in creationism, would its people accept such a way of life? Sex is not all that revered there. In really drastic times, drastic measures are sometimes taken, and China has a way of taking such action.

Other highly populated countries, such as India and some African nations, may not want to take such action because of religion. Their present trend of breeding and starvation may continue until outside pressure or the desire to improve their standard of living turns them more and more toward biotechnology for answers.

Self-preservation and, more than anything, prosperity and fanatic desire to retain it will make some members of civilization do things few people today thought would ever happen.

The "electronic drug pusher" will play its part. This is what I call television, because of how it has addicted the world. Governments may come and go, but television is here to stay. More and more people will want a piece of the pie as they increasingly view television.

We are trying to achieve a better quality of life, with pollution control and reduced environmental degradation.

How do you achieve a better quality of life with an increasing quantity of defective children, through no fault of their own, being born and polluting the species? What better example of taking one step forward and sliding two back?

When I call this to the attention of the environmentalists, I get that hangdog look. I tell them forty years ago concern for the environment only got a self-assured, arrogant response from most people, who said there was nothing to worry about.

Aren't environmentalists today environmentalists because of what has happened over the last forty years? Yet they are reacting the same way now, in a sense, to pollution of the species resulting from contamination of female wombs as people reacted to concern for the environment forty years ago.

Nothing to worry about regarding an artificial womb, they say. Long way off. Will never happen.

As a 1970s trend report on genetic engineering put out by the Life Insurance Institute asks: "Is it good and right to seek a superior race via computers, genetics, and behavior modification? Since many questions remain unanswered, the time is well upon us to seek the answers and insure that the best technologies have to offer is not offset by the worst."

Where are the environmentalists and organizations such as Zero Population Growth when we need them? Why aren't they leading profound discussions in this area of an artificial womb for instance? And they have the audacity to accuse the rest of the establishment of apathy and noninvolvement.

Like it or not, we are in that period of history when if we fail to react to the new technologies in biotechnology now being forced upon us, we will cause people of the future to judge us harshly.

Another response I get from listeners when I'm a guest on a talk show is: "We shouldn't play God." The truth is, medical science plays God every day. Society applauds its accomplishments. The person or persons finding a cure for AIDS will win a Nobel Prize.

From pills to relieve a headache to medication for corns on your feet, medical science played a role. There are antibiotics to cure an infection, insulin for diabetes, and organ transplants. Is this playing God, too?

On a regular basis now, we hear on television and read in the newspapers of another breakthrough in genetic engineering that could or will cure a disease or solve some problem. But most people don't realize where this is all taking us. Or is it that they really don't want to know?

Some say we shouldn't play God, while others say God is working through man or woman to make a better human being. A Catholic moralist was quoted as saying, "If man can breed better races by genetic engineering or make babies in test tubes, why shouldn't he do it?"

For those who say we shouldn't play God, am I to understand you prefer we produce defective children for all the reasons I've listed? Even if a better way is found to produce children, including perfection of an artificial womb?

For those who believe we should give free rein to science, do you include crossing humans with animals, as has been mentioned in government hearings?

"God won't let it happen!" I heard that about going to the moon. One hundred years ago, there were people saying, "If God meant for man to fly, he'd given him wings."

Well, he did. But in a way few people then could comprehend. Few people today can comprehend how humans will be produced in the future.

It is said that the more things change, the more they stay the same. The way people thought about the future in the past is the way people still think today. What goes around comes around.

"Can't stop progress" is still a favorite expression. Progress is like crossing a muddy field. As you pull one foot out it only helps to sink the other foot deeper. Pull that foot out and the other one sinks deeper. You'd better keep going if you don't want to be left stranded or sink altogether.

Progress gave us the airplane. Airplanes do crash. Progress gave us life-support systems, which can keep people alive until they are cured but can also keep people more dead than alive.

The few people who filled out living wills twenty and thirty years ago were considered morbid. Now hospitals to protect themselves request patients to use living wills. Even people opposed to euthanasia have had to reconsider.

There are so-called experts who claim we will have plenty of time to think about and discuss the issues and questions that you're reading on these pages. Plenty of lead time, they say.

I disagree. I believe that while muddling through got us by in the past—if you think the way we handled the environment in the past was satisfactory—in this present high-tech era we must get out in front just to keep from being overwhelmed. Unless you feel all leaky roofs should be fixed only after the rafters have rotted.

The artificial womb should already be part of the American dialogue. We should not wait until it's announced on the evening news by some anchorperson.

Or would you want to wait until you're standing in line waiting for gas, as you did during the energy crisis of the 1970s, then have everyone pointing fingers of accusation at each other for being apathetic and lacking involvement earlier? If you defend the position that nothing can be done until it has happened, then isn't all this accusation of apathy and noninvolvement on the part of the public plain hypocrisy?

Let me remind you again of the following statement in the life insurance industry's trend report on genetics: "Since many questions remain unanswered, the time is well upon us to seek the answers and insure that the best technologies have to offer is not offset by the worst."

I often wondered what the pro- and antiabortionists think about an artificial womb. When asked, they, too, usually answer that it will never happen or it's a long way off. One participant in a discussion on abortion when asked replied, "Once we get people used to abortion, we should have no trouble later getting them to accept an artificial womb."

My response to abortion is that it is only a temporary phase we are going through on the way to perfecting an artificial womb. In years to come—the middle of the next century?—people will wonder why there was all the hassle during this period.

Once the number of defective discards from an artificial womb no longer exceeds that of spontaneous abortions now taking place in pregnant women, then an artificial womb will be accepted.

Certainly the millions of abortions performed worldwide each year don't help the argument to retain pregnancy in women, if a better way is found to produce life. Discarding defective embryos in a laboratory may prove to be far less barbaric than ripping and sucking those millions of fetuses from the wombs of women, often only to be thrown into garbage cans. Or will some of those fetuses now used for fetal research to cure diseases and defects be used more to help perfect an artificial womb?

At least one would think in laboratory births the potential humans would be planned and wanted. No need to abort because of inconvenience, the child being the wrong sex, the mother being a victim of rape or incest, or the pregnancy being a plain accident and the child unwanted.

Unwed teenagers' pregnancies are causing plenty of restless, angry taxpayers. These teen pregnancies and the accompanying welfare cost billions of dollars, taken from middle-class Americans, are causing many to see red on their economic balance sheets. That pocketbook nerve is hurting.

Add to that the moral and ethical implications and consequences. With thousands of babies now coming from contaminated female wombs, plus thousands more of unwanted pregnancies and abandoned babies, eventually those implications and consequences can no longer be ignored.

If there are those in American society who strongly endorse the right of a woman to have control over her body, what happens when another way is found to produce children other than use of the present female uterus? As some members of the feminist movement have been quoted saying, "Science should be used to release women from biological reproduction!" Will "penis envy" become "womb rejection"?

I remember attending a symposium in the fall of 1973, just when the Arab oil embargo began, on the subject of genetic engineering. One of the speakers was the president of NOW (the National Organization for Women). In her address she kept using examples of how society felt women could not compete and be compared to men. Each was a putdown of women, and after every example she would say, "But can men conceive!"

Little did she realize that within five years, yes, men in white coats under bright fluorescent lights in a laboratory could conceive human life. What she so haughtily thought was an impossibility became a reality.

At the end of the symposium, the slate of speakers invited questions from the audience. Counting on just such a possibility, I had prepared my questions, neatly typing them out before I arrived at the symposium.

When the moderator of the panel came to my typewritten questions, all he said was, "These questions are loaded," and they were ignored, never read out loud to the audience or submitted to the panel for discussion.

One of the questions I asked was: "Is it not true that any

day now a virgin or woman sterile due to blocked fallopian tubes could give birth to a child by the means of in-vitro fertilization?"

This took place in an institution of higher learning. The chancellor of the university at that time became governor of the state four years later.

Here is another excerpt from that Life Insurance Institute report on genetic engineering:

> If we have the capability, why shouldn't we upgrade the health, life span and functioning capacity of those who inherit the earth? Why shouldn't women be liberated from nine months of pregnancy if we can produce healthier children in other ways? Would it not be a great thing if people's mental capacities were developed to their fullest, giving them greater control over themselves and the application to which their knowledge can be put? Is this not a wonderful legacy to leave to our children at a time when we feel increasingly incapable of coping with ourselves in our present institutions?

No, this is not out of some Nazi manifesto. This comes right out of the ivory towers of the American life insurance companies.

When it says "at a time when we feel increasingly incapable of coping with ourselves in our present institutions" am I to understand that the life insurance industry already feels motherhood as we know it today is rapidly coming to an end?

When you hear of so many women who will go to any lengths and means to achieve a pregnancy, you wonder if any woman would gladly give up pregnancy if a better way was found to replace it. Since they resign themselves to the fact that there isn't anything they can do about it, you rarely hear women voice their feelings against pregnancy publicly. It's just not the accepted thing to relate and discuss. Is the big breakthrough in biotechnology about to change all that?

"If men had to produce the babies, there wouldn't be many born" is one statement you hear coming from women occasionally. It's usually said in a tone of voice with some bitterness and resentment.

Wait a minute! Scientists in Australia feel a pregnant male is possible. Technology is already available in embryo transplants and cesarean section. Achieving an abdominal pregnancy is the only missing link.

I brought this to the attention of listeners to a Canadian talk show. A nurse called in shortly after and said they had studied in nursing school the possibility of fat surrounding the male intestines supplying the food needed to feed the fetus.

If husbands are now encouraged to share equally in household chores, what will happen when male pregnancy is a reality? Will each partner in a marriage or live-in partners share in producing the offspring? Or would it be economically advantageous for only one to produce the children, if the other one has a more lucrative career?

Is this a way for homosexual men to have children, just as lesbians now have children through artificial insemination? Women can do that in the privacy of their homes by inseminating male sperm with a turkey baster. For men it would be more involved. A trip to a clinic or hospital to insert the embryo would be needed. But who knows? A simpler way may be found.

The way around this will mean science producing healthier babies than any female uterus or male pregnancy can compete with, especially when that artificial womb is free of smoke, alcohol, drugs, and venereal disease.

"Turning off the baby factory" is said sarcastically by some women. What many women might view as a great gift of creation others may think of as only another assembly-line production.

One woman, who was briefly a U.S. senator, filling her

deceased husband's term, stated in an interview that pregnancy was the penalty and burden women had to pay for the original sin in the Garden of Eden. How many other women think this way but never express it, or at least not often, because they think it's not morally and socially accepted?

Think about this scenario. A television camera focuses in on a happy, smiling woman. In her arms she holds a baby. Then she looks directly into the lens of the camera and says, "This is my baby. I picked him up at the WXYZ Incubator Clinic just this morning. He is guaranteed to be healthy. He was monitored for any abnormalities from the time he was conceived in a petri dish until he was produced from an artificial womb.

"My husband had a vasectomy after he deposited samples of his sperm in the ABC Sperm Bank. I had a tubal ligation after my eggs were deposited at the ABC Ova Bank. The burden of pregnancy and worry over having a healthy baby was no concern of mine."

She added this with a twinkle in her eye: "Yesterday I went waterskiing. Later my husband and I dined at our country club. When we arrived home, we enjoyed sexual intercourse.

"No stretch marks. No torn vagina from a difficult delivery. I highly recommend WXYZ Incubator Clinic when you consider having your next child."

The television camera then turns and focuses in on another woman. She appears unhappy and forlorn. She speaks:

"I'm two weeks' overdue. I've had morning sickness for six months of my pregnancy. Two of my four children have genetic defects. As a result of amniocentesis, I've been told the child I'm carrying is normal. That still hasn't eased my mind.

"And then there is that weight problem afterward. It's plain torture for me to lose that weight I gained during pregnancy. The last months of pregnancy and until I lose the excess weight after pregnancy I know I'm not attractive to my husband. Perhaps that's his problem, not mine. Still, it hurts."

If they had their choice, which one would women choose? There are women living today who will see this become a reality. When do the debate and discussion as to its impact begin?

As a former energy czar said, "In a democracy, people don't want to face up to a crisis until they are right up against it."

An editorial in a Chicago newspaper expressed concern over the increased number of cesarean sections being performed. It stated that maternal mortality was 40 per 100,000 for cesarean births compared to 10 per 100,000 for vaginal deliveries.

During still another radio talk show I was a guest on, a young man called in and said if he could pick up his child from a laboratory free of many birth and genetic defects and his wife didn't have to go through the risk of childbirth, he was all for it.

A minority of one? How many more out there feel that way? Why is this being suppressed when it seems nothing else in our society is taboo?

Some young women observing for the first time the birth of a baby either faint or come close to it. The fear of pregnancy and its necessary consequences may be more widespread in the female population than we realize. How will they react if another alternative to pregnancy is perfected that means no risk and inconvenience for them?

Remember that woman senator I spoke of earlier? She also said in that interview that "a woman is not at her best while giving birth."

Is the breaking of a pregnant woman's amniotic fluid and spilling it on the living room rug and giving birth to a child in the kitchen or bathroom better than a child being delivered from an artificial womb in a laboratory with attendants present under those bright fluorescent lights?

In the future will giving birth to a child in the back of an ambulance racing to a hospital be considered more efficient and acceptable than giving birth in a sanitary laboratory? Can a scene where a woman is screaming with pain while giving birth to a child in a hospital delivery room and cussing some man for getting her into that situation ever be as orderly as the laboratory setting?

Is a life resulting from a man and woman, both intoxicated or high on drugs, having sex in the backseat of an automobile better than a life planned and wanted produced from a laboratory? Or how about a child who is the consequence of a rape, incest, or both? How many present-day pregnancies result from these and other similar experiences?

Next time you think about the national debt, balancing the budget, and paying your income tax, also think about those things. Your pocketbook is being hit by more than just crooked politicians.

Biotechnology can and will change social conditions. It will in the process change how we now look at sacred and cherished institutions and traditions and perhaps eliminate them. We adapt. Just as we adapted from a horse and buggy age to expressways with a minimum speed of forty-five miles per hour.

When you compare our present modern technology in transportation, communications, and electronics, for example, you have to question just how efficient our present means of human reproduction are in comparison, even if childbirth is still considered a miracle. Splitting the atom and breaking the sound barrier and gravitational barrier in space exploration are miracles, too.

Do I hear that apples-and-oranges comparison again? I've told you biotechnology is going to make fruit salad out of those apples and oranges.

You say it still takes God to make a tree. God may make

a tree, but scientists through cloning can make acres of those trees all look the same. The timber industry feels this is to its advantage. Money or morals?

Already lawsuits and huge increases in malpractice insurance premiums are driving gynecologists and obstetricians out of the profession, because people want perfect babies. Procedures such as amniocentesis, fetoscopy, and ultrasound, just to name a few, are placing an added responsibility on attending physicians.

What about the bonding of a child and its mother during pregnancy if the mother's womb is replaced by an artificial womb? Has lack of a mother handicapped orphans? Has it hindered adopted babies?

Maternal instinct? Conditioning or instinct? The way television can put the ring in the nose of people and lead them from fad to fad, the maternal instinct should only offer a little tougher challenge to those white-collar—or is it blue-jean now?—executive producers.

When I ask scientists if they really want to make us less human, some answer, "Yes, if it makes people more humane."

Let society paint itself far enough into a corner, and some bizarre and desperate measures may be resorted to in order for society to get itself out of that corner.

Forty years ago, I never would have said what I'm about to say: women are far superior to and more resourceful than men! There is no doubt in my mind that it takes more courage to be a woman than a man. Polls show most women and men agree it's more difficult to be a woman.

To top it off, through all their difficulties, women outlive men. The idea that women are the weaker sex is pure myth. That so many remain so cheerful through all they endure amazes me. Take pregnancy, for instance. A belly pregnancy simulator has been designed to give males and especially single teenage girls an idea of what the latter stages of pregnancy are like. It's no fun. Thirty-five pounds of water, lead balls, and buckshot are strapped to the abdominal area. Then the person is told to do household chores and try to sleep. The participants soon discover there is not much romance to having babies. Remember, too, this simulator doesn't include morning sickness and, in the end, the labor pains.

No wonder some women cuss out their husbands in the delivery room. Some men admitted, after being present while their wife was giving birth to their child, that if they were women they would kick their husband in the groin every time he wanted sex.

It always amuses me that on a Sunday afternoon during the football season some 250-pound linebacker wants to prove his power, his brute strength, but stack him up against a woman half his weight giving birth to a child and see what a screaming meemie he would turn out to be if the tables were turned.

Then take menstruation. That alone would turn a man into a real bitch.

At a young age, a girl experiences her first menstruation. For the next thirty or thirty-five years, more or less, she may be one of the unfortunate and endure a living hell each month. Having a heavy flow, causing us to change sanitary napkins more than six times a day, could make anyone irritable.

Even including those more fortunate women, most women will agree menstruation is still a curse. In some primitive societies, women are isolated during this period. Few women, if any, will defend menstruation as they would pregnancy.

I've backpacked with women above the Arctic Circle in Alaska. They weighed 100–125 pounds, their packs more than 60 pounds. They crisscrossed streams and rivers in water up over their knees, water from ice melting and flowing into those streams and rivers from glaciers high above us in the mountains. Some I knew had to be experiencing menstruation. If they had cramps, pain and/or bloating, I never knew it. They always had smiles on their faces. And you tell me women are weaker than men. Try it on someone else. To top it off, when in that condition, because of the blood, women are more vulnerable to grizzly bear attack than we men. Everything works against them. And then in some societies, male babies are preferred, female babies rejected. No wonder so many women resent men.

Later comes menopause. Along with hot flashes, there comes possible pelvic relaxation, with its tipped uterus or a dropped uterus and all its implications. You would think years of menstruation would be enough.

Men may later in life have concern with prostate problems. A woman, because of the reproductive tract, most of her life has a continuous concern about breast cancer and vaginal and urinary infections. She has regular cervix and uterine examinations that can keep her on the mental edge of worry. It takes a different mindset than that of men to endure a periodic probe such as a Pap smear, which might feel worse than having a tooth drilled.

And if that wasn't enough, from birth to death, a female soon learns she must keep one eye looking over her shoulder for that sex abuser and rapist.

It's a rare man indeed that would change places with a woman after realizing just what she goes through in a lifetime. Oh, I hear occasionally a male make the derogatory remark that he would spend all his life lying on his back if he were a woman.

"It's a man's world!" At times you hear this said by women, usually said if not with resentment, then with resignation.

Across the radio waves comes the sound of a newborn baby crying. Then a woman's voice says, "This baby was born handicapped. It was born female." This was once a public announcement promoted by the feminist movement.

When I addressed a mixed audience of men and women and asked the women if they really wanted to give up the joys of motherhood, one woman practically shouted at me, "And what makes you think it's a joy to be a mother?"

It wasn't this woman's reaction that startled me, but the attitude of the rest of the women in the audience. They looked at me and actually defied me to rebut her charge.

When I addressed another audience of only women, I asked them if they would care to become pregnant in order to become mothers if science could replace biological reproduction with laboratory reproduction. The reaction was mixed.

Then I came up with the punch line. I asked them if they would still want to become pregnant and be mothers if science could replace biological reproduction with laboratory reproduction and also include eliminating menstruation.

The reaction was like air suddenly escaping from a bag. The women became subdued and slowly shook their heads no. Some had a look on their face and hung their heads in a way that projected a degree of despondence and resignation. It was a look of finally admitting that the cherished institution of motherhood could not survive even that.

Women are gaining more equality. "Equal pay for equal work" is only the tip of the iceberg.

Only in a gender-free society will women get true equality. Is civilization moving in that direction? Will more planetary exploration bring this evolution about?

For women it will be easy. The urethra is there. The rest could become as useless as the appendix.

As for men, I've jokingly said, "Why should men carry around a chandelier when a single light bulb will do?"

I've watched society in my life become more regimented with its computers, credit cards, and couch potatoes, and this has been accepted. Regimentation is not considered restrictive by many if increased convenience means more freedom, if it means increasing their standard of living, if it makes them more prosperous. For others, if it means it's necessary for survival.

So it includes both the liberals and conservatives. Deny either one and the howl is heard across the land.

Biotechnology has and will help to keep society on this course of regimentation. Slowly but steadily it is making what is normal, abnormal and what is abnormal, normal. This will continue out of necessity or economics or, more accurately, out of a combination of both. Fear of AIDS may prove to be a determining catalyst.

Absurdity is becoming a reality!

Sex in an Ever-changing World

It is obvious, although few yet agree, that biotechnology will bring motherhood as we know it today to an end. Just when is hard to say. This present course society is following, which is costing the nation billions of dollars, will only hasten the day. That "pocketbook nerve" just keeps on getting more sensitive.

What the end of motherhood will mean for the future of sex is another debatable issue for future generations to face. If we don't find a cure for AIDS soon, the end of motherhood and what happens to sex may be determined much sooner than we think, and especially for those who think some things can never happen.

At one end of the spectrum sex is viewed as glorious and a God-given gift of love, while at the other end of the spectrum sex is looked upon as repulsive and obscene. When technology can succeed in manipulating the hormones and control the sex urge, with few, if any, side effects, which attitude will prevail?

It is surprising how often you already hear sex referred to as an addiction. Has it been reduced to being no different than alcoholism, this once sacred function?

Some celebrities use sex addiction as an excuse when defending their action when they are confronted with a sex charge or scandal. When a popular married athlete confessed to having a mistress away from home on road games, which resulted in a scandal, he blamed it on his addiction to sex.

Periodically one will read a letter to an advice columnist in which the writer complains about sex and what it has done to him personally or what it is doing to society. One such letter from a married woman said that for her sex was only a duty. She asked the columnist to poll her readers as to how many other women felt the same way.

The columnist received 200,000 responses. Half agreed with the woman who wrote.

A married couple wrote to another columnist that they were quite content and happy even though they had dispensed with sex. Again a surprising number of couples responded, of all ages, living with a similar arrangement and satisfied to have it stay that way.

When a woman writes that it would be best if sex was eliminated altogether, the usual answer coming from the columnists is: "Sorry, honey. As long as we continue to make babies, we will have sex."

Wonder what the answer will be when biotechnology finds a different way to produce babies?

It has been said that sex is a dirty trick God played on mankind—pardon me, personkind—in order to perpetuate the species.

Lord Chesterfield, an English statesman and author in the 1700s, described sex as follows: "The pleasure is momentary, the position ridiculous, and the expense is damnable."

Sex is a momentary pleasure? For both sexes?

How many wives find themselves frustrated and unsatisfied after sex—or should we use the modern terminology: *making love*—while their husbands have already rolled over and gone sound asleep? How often has this been a cause for an appointment with a marriage counselor?

For men sex can mean seconds. For women it means minutes. This can cause much anger between the sexes. One of the reasons for the "war between the sexes"? Sometimes this means men are moved to the "rear-echelon area": if not to another bedroom, then to the couch.

Some women, like the married woman who endures sex only as a duty, find not even momentary pleasure in sex. While many women may enjoy sex as much as men, if all women thought as often and as much about sex as men, we would have a nation of nymphomaniacs. This might be appealing to men, although even the present-day increased sexual aggressiveness in women is having its repercussions.

As for the positions being ridiculous, sex cannot be called a dignified performance no matter what position is chosen. Maybe that is why sex is so popular a subject for comedians. Are embarrassment and humor that far apart?

Makes one wonder how an attractive, well-dressed, elegant, and sophisticated woman passing by can also be a call girl and turn to prostitution. Is it love? Is it hormones so powerful it can make this woman offer to customers a wide variety of perversions that modern society is trying to make normal for all? Or is it even here that lust for money can win out over reason?

A poll was taken as to what was more important, cars, microwave ovens, or sex, Guess what? Sex came in third!

I've always known automobiles have become the bitch goddess of the American public. But microwave ovens. Wow! Materialism just might win out over sex. If this poll is accurate, it might have already.

How about Lord Chesterfield and his claim about the expense of sex being so damnable?

Books have been written about the pleasures of sex, about the positions of sex. Those might be the titillating end of the sex spectrum. But it's the other end of the sex spectrum, with its suffering and anguish, that has filled many more volumes. It is that end of the spectrum that may determine the survival of civilization.

One picks up a daily newspaper. On one page it reports a year-old female being sexually abused. Turn the page and another article relates how a ninety-year-old woman was raped in a nursing home. You call ours a civilized nation?

These are rare exceptions I used as examples, you say? There are enough women between those ages that my examples are no longer the exception, enough women and even males being sexually abused, assaulted, and raped each day, so that it is becoming commonplace.

Why must women have to keep looking over their shoulders from birth to death for that sex abuser and rapist? Just how free are women to enjoy the things men take for granted? A man alone may be in danger and fearful, but a woman alone feels more fearful and experiences more danger way out of proportion to what the man alone feels.

Women are bolting their doors, carrying Mace in their purses, and taking courses in self-defense. As women lawmakers approach the majority in government, will changes now unimaginable take place? How long before women have had enough of being sex objects?

One leading member of the feminist movement when asked when women would finally have equality replied, "In about one hundred and fifty years if we're lucky."

Maybe much sooner than that. Knock a hundred years off of that 150-year estimate. That juggernaut of the scientific/medical/industrial complex already has society reeling, and this is just the beginning.

I've asked why women should have to keep looking over their shoulders from birth to death for that sex abuser and rapist. Now I ask: do we have the right to deny women freedom from fear of sex abuse and rape when the sex urge can be brought under control? That control will be achieved. Now it's in the infant stage. Biotechnology will bring about just such an achievement.

There is a human condition called diabetes insipidus. It's called diabetes because it is similar to that disease, with intense thirst and excretion of large amounts of urine as symptoms. It is a disorder of the pituitary gland and not blood sugar. The condition can be brought under control by squirting a spray up the nose to make certain hormones function properly.

The pituitary gland and the way it is manipulated will play a major role in making what is normal, abnormal. Already it is being used to increase height in people such as dwarfs. Some athletes use steroids to manipulate hormones to increase their strength.

While some research in that field is now abused, when it is finally refined and causes little or no side effects will it then be accepted? Is this the way science through biotechnology will bring the sex urge under control and even eliminate menstruation?

Take menstruation. Why should women have to put up with such things as premenstrual syndrome and toxic shock syndrome when menstruation and its side effects each month could be eliminated with a mere squirt of a spray up one nostril?

Again I ask, how many men would want to be subjected to changing sanitary napkins more than a half a dozen times a day during their menstrual period as happens with some women? Or the use of tampons, which can threaten your life through toxic shock syndrome?

Doesn't this "curse" only add to the resentment women have toward men? That expression "it's a man's world," when said by women, is uttered with resignation and not joy in this instance.

Back in the 1960s, a drug was injected every three months into women on welfare who couldn't read directions for taking the pill as a contraceptive. It not only stopped pregnancy, but after eighteen months also stopped menstruation.

This was then found to be convenient to give to mentally retarded women in institutions. It was discontinued when it was thought to cause cancer. Recently it was allowed back onto the American market. Now it is supposed to actually stop some forms of cancer in women.

What happens when a drug company can refine a way where a sniff up the nose of a spray can be a foolproof contraceptive and eliminate menstruation, too? Will the sanitary napkin and tampon manufacturers oppose it because it's a threat to their profits, just as oil companies now oppose alternative energy sources?

Menstruation is a normal function and shouldn't be tampered with, you say? Did that stop the birth control pill from being accepted? Ban the pill and it would cause a worldwide revolution.

True, all women can't use the pill. Other forms of contraceptives keep coming on the market to add to existing means of preventing pregnancy.

As I write this, a new drug taken to cause early abortions, similar to a miscarriage that is a mere speck in the toilet bowl, is controversial. It's already accepted in some parts of the world and will be in ours, either over the counter or under the counter.

The same will happen with the elimination of menstruation—again a case of biotechnology making what is considered normal, abnormal.

If a similar breakthrough in biotechnology can control sex with a sniff of a spray up the nose, can this be regarded as progress or regression? Certainly with millions and millions of women each year being sexually abused, how can we call this abuse civil and progress? But normal, you say?

In the case of rape and sexual abuse, would accepted control of the sex urge be a boon to women who must now live from birth to death defensively as a result of this fear? You would rather we continue on this course of fear, even if a humane method and one not harmful to health to curb sexual aggression is achieved? How many will feel that way when it is a reality?

When a rapist is loose on a college campus, it is the women who are put under a curfew—the typical example of the victim being penalized. How often are the males in the area quarantined? Why aren't the men put under a curfew, with female police patrolling the campus? Is it right that women, because they were born female, should be made to suffer? Just another reason for women to resent men.

It is recommended women carry condoms in order to have so-called safe sex. Will people in the future look upon this as one big joke, which society now takes seriously? How far into the future?

Already this "big joke" has had a tragic consequence. A woman about to be raped by a stranger who entered her apartment asked him to use a condom because of her fear of AIDS. Her action was enough to allow the rapist to be pronounced "not guilty" in court. The justice system came to the conclusion she had consented to sexual intercourse and that it was not rape, another example of how women get the short end of the stick when bringing rape charges to court. Thousands upon thousands of rapes are never reported each year, because women know if they report the rape the humiliation they must go through won't be worth the effort.

Safe sex! The methods suggested are often ridiculous. Condoms have been proven to have one of the highest failure rates when used as a contraceptive. Yet the period when a woman is fertile is so short that use of a condom always as a means of having "safe sex" is useless. That being the case, frequent intercourse using condoms during a woman's infertile period would also increase the chance of both partner's exposure to the AIDS virus because of condom failure. It's like playing Russian roulette using more than one bullet in a revolver.

Society knows the only sure way is abstinence. What happens when biotechnology can achieve that in a truly safe and convenient way?

Already because of the fear of falling victim to AIDS some people are resorting to celibacy. If AIDS develops into a full-blown plague, will society encourage, even plead for, the technologists to curb the sex drive, just as society is now turning to them to curb pollution?

Many no doubt are praying for an AIDS cure soon, so they can continue on this course of "if it feels good, do it," so they can continue having pleasures of the flesh without a guilt complex.

Now that sex is being classified as an addiction, we know halting an addiction is not easy, especially in a prosperous and affluent nation such as ours, where self-discipline and responsibility are only for the other person. Or for "nerds," as they are referred to by the pop culture.

Date rape has become a national crisis. Would a spray sniffed up the nose to curb sex for a certain length of time be accepted when it becomes available? Is this what women would carry in their purses to replace those "safe sex condoms" and to know for sure they would not be raped?

You may laugh and roll your eyes at such a possibility. Remember, biotechnology has the ability to make what was once abnormal, normal. Menopausal women are having babies in their fifties and older, achieved by in-vitro fertilization. Will that become normal as it becomes more frequent?

I don't know what your definition of *normal* is, but here is one definition that covers a wide range: "Normal is the highest rate of frequency in a given environment."

That being the case, we have groups in our midst now that can even make insanity seem normal. On the other hand, what was once considered normal, today is considered idiocy. Bloodletting, once accepted as a medical procedure, now is looked upon as foolish. Life-styles and time have a way of changing abnormal to normal, and vice versa. Will what we consider presently as sacred go the same way?

There was a time when the idea of a new car costing $2,000 was outrageous and people felt that prices could never keep on increasing because no one would buy cars. Now if people can buy a new car for $10,000 they often consider it a good and wise purchase. While this may be a poor example to compare to motherhood, the family unit, and sex, have I made a point? Again let me remind you, the covered wagon was once the normal and accepted way of travel.

I mentioned earlier that a drug available as a contraceptive can eliminate menstruation, too. This drug can also be used as a form of chemical castration and is prescribed in some states for rapists. It is ironic that an heir to the company that perfected the drug was asked to use it on himself when convicted of sexual abuse.

Some rapists have taken the drug voluntarily, even with cancer as a possible side effect. When asked why, one rapist replied, "I'll take my chances with cancer, rather than continue to go through hell with this uncontrollable sex drive."

In the future, will that scientific/medical/industrial complex I speak of be able to detect potential rapists even before puberty? Can a substance, similar to insulin for diabetes, then be given to control that abnormal sex drive?

How much would this decrease the prison population and thus reduce social costs, both morally and financially? Presently abnormal sex drives are costing this country billions of dollars.

Still concerned about that national debt you're leaving your children and your grandchildren? Still concerned about them not achieving the American Dream and maintaining your standard of living? In the end, again, economics will play a determining role.

Some claim rape is not the result of sex, but rather an act of violence. Show a man dying of thirst some water, and won't he try to kill for a drink if denied one? Aren't we doing the same in our present environment with this sex obsession on television, in movies, and in pornography? Aren't they guilty of rubbing raw and bringing to the surface the hormones of young men with this obsession? To say what these money-making exploiters are doing shouldn't bother healthy, normal men is a cop-out. Showing movies and presenting television programs whose theme is how to maneuver women into situations beyond the control of these women and ending in rape is what is not healthy and normal.

Rapists, when interviewed, reveal just what part pornography played in their crime. Is the information suppressed in order to protect some money-making business? What we won't do to add to the Gross National Product! Those prisons to maintain and house those rapists add to the Gross National Product, too, but in quite a detrimental way.

How can we call ourselves civilized when women are so degraded by this undesirable pornography available in our society? The feminist movement has a legitimate right to object to this treatment. The hiding behind the First Amendment, along with AIDS, may force technology to produce something to curb the sex drive, so as to bring some sort of stability to a society slipping into chaos.

The First Amendment may have its merit, but carried to extremes it can become a tyrant comparable to a despot. Our founding fathers never thought it would turn out this way. Is this progress?

Then there is sex harassment. With sex harassment accusations becoming more frequent, men may consider administering, when available, something chemically to protect themselves. The finger pointing of sex harassment at a man by a woman can mean his losing his lucrative job. It can mean the end of his career.

What happens when the charge of harassment is only a form of revenge? How can a man protect himself? Will biotechnology play a part in the solution?

The relationship is becoming more and more difficult between male and female, and especially for the male wondering what course to follow. Many of the advances made by men in the past are cause enough now to be subjected to a sex harassment charge.

Men can get a negative response from just opening doors for some women. A second thought must be given by men thinking about complimenting a woman on her attire, her figure, her attractiveness. How a man reacts to a woman who flirts could mean, legally, his coup de grace. Wonder how a society of neuters would treat each other?

How might homosexuals react to an accepted drug-induced sexual curb if they thought it could improve their acceptance into society? Some homosexuals take pride in and even boast of their sexual orientation. Still others, because of the hell it causes them, turn to suicide.

Many, if not most, look upon sodomy as repulsive, a perversion. Will future generations look upon it as the worst form of barbarism and debauchery?

What will the attitude toward sex continue to be? Will liberalism carry it so far that public sexual intercourse will be as common as hugging on a park bench or kissing your wife good-bye at the airport? Or will the pendulum swing back to where few today can comprehend the extent of the extremity of its arc?

It is not uncommon to hear or see an entertainer getting applause and hilarious reactions to his sexual jokes and actions. One has to wonder just how many in the audience spent a recent Sunday morning listening to a clergyman tell them what a glorious gift from God sex is and how sacred it is. The hypocrisy and double standard of sex in our society can only be complete confusion to any alien from outer space as to what to say and do regarding sex.

It is interesting to observe how people react to a hard-core pornographic movie. Some are immediately disgusted and walk out. Those that stay often react with laughter, giggling, and embarrassment. I assure you few will react as though they were viewing sunsets over the Grand Canyon. Unless you think titillation and inspiration are the same.

When the hormones rage, anything goes. Could this be the only explanation for having oral sex with yeast-infected female genitals or gonorrhea-diseased male genitals? This our modern society calls "making love."

Passion and hormones are the only reason men have sex with some women. It certainly isn't love. There are men who relate in derogatory terms their sexual experiences with women.

"Put a bag over her head" is an expression men will use when advising another male hesitant about having sex with an unattractive woman. As one man told me, "It was like having sex with a pile of warm dung in a barnyard."

Women relate, too, their experience with having sex with some men. One woman said, "It was like having sex with a beer-smelling ape."

Maybe that is why so many men and women must resort to fantasy during intercourse. Only closing their eyes and imagining this is all happening with some sexy celebrity helps them to enjoy sex. A Hollywood sexpot has probably saved many a marriage.

Was it love or passion, for instance, that enabled a prostitute in Honolulu during World War II to service some three hundred military personnel, with help from their hormones, in less than one twenty-four-hour period? If a segment of society finds such action repulsive, what happens when the sex craving can be controlled? Will it be judged then the same as we view alcoholism now?

A glass of wine before a meal is enjoyed and accepted. But drinking a bottle of that wine can result in a drunk driving down a highway and slaughtering a whole innocent family. An alcoholic knows only abstinence is the cure.

Our obsession with sex is no different. It's out of control. As Lord Chesterfield would say, "the expense is damnable." The expense economically, physically, and mentally is truly getting damnable.

That it is still necessary to perpetuate the species perhaps does wring some truth out of the belief that sex was a "dirty trick" God played on humankind. Adam and Eve never thought it would turn out like this just from eating that forbidden fruit. In my life thus far, I've observed how the biblical command to "go forth and multiply" has been challenged because of overpopulation. What happens when technology can change our perspective toward both birth and sex?

The Greek philosopher Plato was supposed to have said that old age finally frees one from the slavery of one's passion. Only passion and the flowing of those hormones from point A to point B could be the explanation for why the most unattractive part of the human body is the most popular. It all starts in the head, and that's where technology will start manipulating those hormones.

I've been blunt in my questions and statements, as blunt as professionals in the health field warning about AIDS. They will become even more frank as the fear of AIDS gets greater and other diseases result from its spread. As the advice over how to practice safe sex with condoms becomes an even bigger joke, the debate and discussion will become ever more heated.

If the things I've written are disturbing and offensive, is it because they smack too much of the truth? Something we don't want to face? You would rather we live in this dreamworld of irrelevant and trivial media information? If you challenge what you've read so far, why not direct it at the media messengers: radio, newspapers, television? But do you really want to know the truth?

The high-tech age we're in won't allow us to stick our heads in the sand and maintain the luxury of remaining ignorant. Biotechnology and genetic engineering are more than just DNA, lowering the frost level of strawberries, or using a hormone to increase milk production in cows. They also can mean eliminating motherhood and controlling sex.

I am concerned and resent the slanting of biotechnological information. Few would be against the announcement of another biotech breakthrough curing a disease or solving some medical problem, but let's tell the whole story.

Science needs better relations with the public and to cease its arrogant elitism. It's tampering not only with lives, but with the whole social structure. It, along with the media, should be encouraging dialogue on the whole ball of wax and not suppression and censorship of certain phases.

When? When does the dialogue begin?

The Family Unit

The family unit is going the way of the family farm! Fifty years ago, if you told those farmers then, in their bib overalls and straw hats, that farming would be the operation and business it is today, they would have said, "Sonny, you're drinking some mighty strong liquor."

The family farm wasn't meant to survive. It was destined to be sacrificed to America's affluence, higher standard of living, and technological progress. Americans didn't want it any different, and that includes most farmers.

I was reared on a farm. I remember walking behind a team of horses pulling a piece of equipment so antiquated now, it's in some museum as a novelty. Tractors costing $100,000, more or less, have replaced those horses. These vehicles come equipped with all the gadgets imaginable: electric rearview mirrors, stereos that give you the best in music while sitting in a comfortable cab, and hydraulic seats, replacing burlap bags stuffed with straw as cushions.

Years ago, even on the hottest summer days you kept your shirt sleeves rolled down while shocking bundles of grain to stop the pricking of thistles. Grain is harvested now from air-conditioned combines. An old-fashioned threshing machine will stop cars along the highway just so the people can watch it operate.

You took a few minutes under a nearby tree when hoeing in a cornfield to prevent heatstroke. Now a herbicide tank takes care of the weeds.

I remember sitting on a homemade wooden three-legged

stool milking cows until my arms ached. Today the cows are milked in a place called a milk parlor. Those ten- and twenty-cow herds of old have increased to 100- to 1,000-cow herds, milked not just morning and night, but around the clock.

An instance was related to me as to what happened on one of these large farms. It seems a cow somehow got down into the area where people attached the milk machines to the udders of the cows. The farmer stopped everything, went and got a gun, and shot the cow.

Sacrificing one cow cost a lot less than having it go on a rampage and destroy thousands of dollars' worth of milking equipment. On small family farms, cows were given names. Today they are just extensions of the milking machine.

Biotechnology has played its part in cows' giving increased milk production. When I told farmers twenty years ago that one cow would soon be having not one or two calves a year but hundreds, I was met with skepticism and even laughter. Now a prize cow through a process called embryo transplants and having other cows carry those embryos to term can produce a multitude of offspring in one year.

Those resulting calves grow into cows with record-breaking milk production, from udders so large they drag on the ground. The contented cow of old grazing in that serene pasture of the past has become just another burned-out victim of progress. And still science wants to inject into cows a bovine hormone to produce even more milk from cows.

Improved technology cut the small farmer's work in half. So what did he do? Go fishing? Not on your life! He got bigger. He and Farmer Jones either rented or bought land from Farmer Brown. Farmer Brown and his family moved to town to help produce all those goodies Americans were demanding. Economic attrition squeezed out some farmers. Their paychecks and plastic credit cards helped to fuel the economic boom.

Jokes about the farmer's daughter and the traveling salesman came to an end. Those farmers' daughters went to town and got good-paying jobs or a college education. Some who in the past would have been lucky if they traveled a hundred miles from home soon were traveling halfway around the world and made that traveling salesman look like a hayseed instead. Rarely is the term *old maid* heard anymore for single women.

The farmers' sons followed their sisters to town. This left fathers selling out at retirement, farm auctions everywhere.

Forty-five years ago, a farmer I knew owned a forty-acre farm on the fringes of a large city. He sold twenty acres. It became a housing development. He told me that soon housewives were cussing and throwing stones at him for holding up progress. They were sick of the smell of manure and the sight of his less-than-modern buildings. They wanted to see a shopping center on his other twenty acres.

So more farmers were sacrificed to economic attrition resulting from urban sprawl, with its housing projects, shopping malls, fast-food joints, modern highways, and other American opulence. Many farmers were right there throwing their milk stools, hoes, pitchforks, and wheelbarrels on the bonfire of antiquity as the juggernaut of prosperity kept rolling on.

With the decrease in the family farms came a corresponding deterioration of the family unit. The traditional family of a husband carrying a lunch bucket to work and a stay-at-home wife caring for the children began to become as antiquated as the farmer's hoe and milk stool. Such families are now reported to number less than 15 percent of all families.

Perhaps the family unit wasn't meant to survive, along with the family farm. It, too, was destined to be sacrificed to American affluence, the higher standard of living, and technological progress. The family farm of earlier years, with its large families of sons and daughters to help with the chores, was more conducive to family life.

As families became more urbanized, they, too, eventually became victims of technology. Modern gadgets began to make life easier for the housewives and mothers. Out went the washtub, washboard, ironing board, and clothesline through the back door. In came the automatic washer and dryer and perma-press clothes through the front door. With mothers out and microwave ovens in, fewer and fewer families were sitting down to eat the family meal together. Convenience foods and those ovens put everyone on the go.

Instead of becoming larger, as the farms did, the urban family became smaller, having fewer children. The cost of rearing children was discouraging large families, especially the cost of education.

A college education was no longer for only the few; it was for everyone. That cost money. Another luxury became a necessity.

I observed my generation after World War II. That generation had gone through a depression and a world war. When the war was over, they felt they had suffered enough, so they went on one big binge.

Madison Avenue set them in their favorite easy chairs and placed an idiot box, television, that masturbator of the eyeballs, in front of them. It put a bowl of popcorn in their lap and a ring in their nose. Then reduced them into becoming the "silent majority," a label they took pride in maintaining.

One goal my generation believed in, which was almost total, was that their children would have it better than they had it. For some reason they felt affluence, bordering on opulence, built character.

With all that time on her hands, because of those labor-saving gadgets, the urban wife went to work for wages outside the home. More money meant more goodies. More goodies meant the need for more money.

The urban family soon was on a treadmill similar to that of the family farm. More land meant the need for more expensive and bigger machinery. More expensive and bigger machinery often meant acquiring more land. When the family farmer couldn't make it into a profitable business and no longer just a farm, he either went into bankruptcy or sometimes performed the ultimate act of desperation: suicide.

When the urban family didn't go into bankruptcy, it took another course: divorce. The divorce rate began to skyrocket with divorce decrees granted for all kinds of reasons.

Another problem precipitated the demise of the urban family. Remember that idiot box that found its way into living rooms in the 1950s? It became a wedge between parents and their children. Soon parents were throwing up their hands in despair as they tried to compete with television and peer pressure in rearing their children.

Having never owned a television set, I was surprised when I visited people's homes and observed how television personalities were addressed on a first-name basis, as though they had become part of the family, often closer than other members of an extended family such as aunts, uncles, and grandparents.

I remember when fathers and older men ridiculed the young men when they started to let their hair grow long. Not long after, the older men were also letting their hair grow longer, when they saw it on their favorite anchorman or television personality and it became the style. Television has tremendous power.

Programs encouraging infidelity and the attitude of "letting it all hang out" helped to add more obstacles to the family unit. Divorce, increased teenage pregnancies, more single-parent families, drugs, ghetto gangs, sexual, physical, and mental abuse, you name it; they all hit the fan, and the family unit started down the tubes.

Sometimes it makes one wonder if maybe all these audience-participation talk shows on the deteriorating American family are consciously or subconsciously getting society ready for its complete collapse.

The alternative? Day care centers will become the most expanding and growing service industry of the future, the McDonald's of child rearing.

Well-paid professionals will rear and educate children, once the function of parents. Out of the artificial wombs right into the day care centers. Sounds more and more like a spinoff of Huxley's *Brave New World,* doesn't it?

How will it be paid for? If people consent to a sum of money being taken from their earnings for Social Security, is there something similar out there in the future to pay for producing and rearing children? Since the cost of raising children is becoming such a burden, will society decide it should be spread out over the whole population?

With many people resenting their tax money going for welfare, will they feel this might be a solution? Just how much will people be willing to have deducted from their annual earnings for the convenience and the end to the burden of motherhood and family rearing of children, which are now accepted and revered institutions? Is this what the insurance trend report on genetic engineering and behavioral modification was subtly referring to?

Never happen in a democracy? In a democracy you let things evolve until the majority thinks it's their idea. That's what consensus politics is all about.

Probably what will determine whether the family survives will be the answer to this question: can the family, as we know it today, rear children to the adults needed for our nation to be competitive in the high-tech global economy that is coming?

That is the most important question, if a society such as ours wants to maintain and increase the standard of living of the individuals in that society, to be sure that they can experience and live the American Dream. Are we ready to bite the bullet?

You may say this sounds worse than socialism, and it certainly isn't capitalism. I open a dictionary and this is how it defines *capitalism*: "An economic system based on the ownership of land, factories, and other means of production by private individuals who compete with one another, using the hired labor of other persons to produce goods and services for a free market for whatever profit may be obtainable."

It says nothing in that definition about a family unit being needed. At the time of this writing, the richest man in America is single, not married. It could be a woman. He or she could have come from an artificial womb when perfected. And he or she could have been reared in a day care center. It happens all the time with orphans in an orphanage.

There seems to be a segment of society already that doesn't believe in marriage and the family unit. Living together out of wedlock used to be called "shacking up." Now it has a more acceptable label. Millions are doing it. They even see counselors, similar to marriage counselors, when they encounter problems.

I remember forty years ago if a couple lived such a life-style they had better keep it secret or take the chance of getting run out of town. Now if someone criticizes such an arrangement he might get run out of town. Hear all those boos on those popular television talk shows when someone defends family values.

Occasionally a slipup happens and a child is born, if an abortion doesn't take place first. Then from time to time, and it seems to happen with more frequency, you hear coming across the news how another live-in lover high on drugs or alcohol killed his female partner's baby because it was crying. What you hear coming from some marriages is comparable or worse.

Those who want to experience the joy of raising a family are being overwhelmed by those not fit to be parents. It's the same old story as with airplane hijackers: the majority are penalized in order to control the minority.

Women contaminating their wombs spoil it for those women wanting to be good mothers. And "do it if it feels good," resulting in AIDS, is spoiling it for those wanting to experience that glorious gift from God called sex.

People may want to enjoy their "fun and games," but it's no fun and games for a woman who has been raped and mutilated. "Is making love worth dying for?" is a public announcement coming across the radio regarding the consequence of AIDS. Crossing the legs and keeping zippers closed is still the safest sex and one of the best contraceptives. Will enough practice it before drastic measures are finally taken?

Grim future? I'm surprised to hear so many young people say they think the world won't last another twenty years. Or that old refrain: "We're in our last days." Thousands or more years from now? Don't hold your breath till then.

Past generations felt and said the same things. Yet we continue on. Some say we are better off; others say worse off. Still, how many would want to go back to the "good old days"?

The future will be no different. Especially when the life span is increased and diseases cured, plus an ever easier and more leisurely life-style.

We are in that period of deciding the tradeoffs and debate and discuss their implications, the dilemmas of technology, disadvantages and advantages, and determine if it's all inevitable. Is this our present responsibility to those that follow?

All sounds so absurd? Have you already forgotten how the reality you're living began being absurd?

Conclusion

While standing before George Washington's tomb at Mount Vernon on a clear fall morning a few years ago, I looked up to see a large commercial jet approaching an airport at Washington, D.C., and wondered how George Washington would react if he suddenly came alive to see a plane flying overhead, where its two or three hundred passengers had just finished eating a meal and had watched a movie earlier and attendants were going up and down the aisle serving them drinks, even diet drinks, if they were concerned about their weight.

That airplane can fly from coast to coast across this country in less time than the day Washington often needed to travel in his horse-pulled coach. Our founding fathers in their wildest dreams could never have imagined we would be living the way we are today. Even that sage Ben Franklin, whose trips to France took weeks and now would only take hours, would be amazed.

Earlier I had visited some Civil War battlefields in the area. Try to convince soldiers who fought on those fields that in less than one hundred years a single bomb could destroy a city, that two such bombs would end a war that went on for nearly four years, similar to the length of theirs. It would have been enough to have had you discharged from either the Union or Confederate army for being weak in the head or placed at the head of a charging column just to get rid of you.

In my travels, I periodically encounter portions of the route Lewis and Clark took to the West in 1804 and 1805. This was less than two hundred years ago. How would they respond if they came alive today and saw how portions of that same area they explored are now subjected to traffic gridlock?

Ruts made from wagon trains on the Santa Fe Trail are still visible in places. People in modern-day automobiles travel in a few minutes a distance that took those early pioneers a whole day to accomplish. For airplanes, mere seconds.

At the end of the day, present-day travelers don't have to stand around a camp fire eating beans and bannock. They can end their day by stopping at some ritzy hotel or motel, shower and then eat in the best of restaurants, or fast-food places if they prefer.

Luxuries have become necessities. In America along with its affluence and credit cards there are no more luxuries; they have become necessities.

You will agree with the examples I've listed above, that people living then would be amazed at what has taken place, much of it in less than two hundred years. If this so-called progress continues—and it will probably accelerate—you reading this today would be even more amazed at what had taken place if you came alive one hundred years from now. Much of it will be the result of genetic engineering and biotechnology.

"Weird." "Bizarre." Is that your response to what you have just read in the preceding pages? What makes you think we aren't presently living in a weird and bizarre way, according to those who went before us? Isn't it all relative?

We might look upon our early pioneers as naive and unworldly in their reactions to the way we live now. Who is to say we wouldn't be judged similarly by those people who live in later generations? What is our responsibility to those generations?

Communications and transportation have made the world smaller with its ever-increasing population. Does our responsibility increase exponentionally along with this progress?

Cope with events as they happen, you say? Then why all this complaining as to what has already happened? Have the environmentalists a right to bellyache?

I suppose you feel my emphasis on economics as the determining factor in the direction society takes in the future is an exaggeration. Call me a cynic, but I happen to agree with that philosopher of old who said, "First food, and then morals." As one unemployed auto worker said when asked if a job was more important than love, "You can't live on love." As the saying goes, "love may make the world go around, but it's money that greases the wheels."

Have you ever counted the mail you receive in a year that isn't related to money? How many grieving widows have to spend as much time worrying about the possible economic deprivation over the loss of a husband as they do feeling the actual sorrow itself?

Money means everything. Will the young prove to be the "Doctor Death" of the old if it might lower Social Security taxes or health costs or prevent a depression? Has the country lost its soul not to the company store, but to shopping malls and video outlets? People kill now not for need, but for want, and show no remorse.

How many evening news programs don't mention money? How many elections are not determined by economics? If love is more important than money, how come couples about to marry sign a prenuptial agreement?

Sex is more important than money? Just how many people would give up sex if it was a choice between either automobiles or microwave ovens and sex? Microwave ovens are still hard to believe. You'd be surprised at how much impotence is caused by worry over economic difficulties.

I often ask people of my generation and older if they would want to go back to the "good old days." I have yet to get a "yes" to my question. I have received a resounding "no" sometimes even before I finished the question. Usually the answer is a hesitant "no" and they add, "But I sure would like to have those simple, more peaceful, and contented times back."

Why? Why do they answer, "No"? Because they know since the end of World War II they have lived a more prosperous and affluent life.

And when those farmers of old complain about what's happening to farming, I say, "Okay, we will give you back those forty acres, a team of horses, a hoe, the milk stool, and a standard of living that life-style will support." I get no response.

Beneficiaries or victims of progress? It was Henry David Thoreau who said men have become tools of their tools. Add women, too.

The majority of people living today in American society were born after World War II. I find it amusing, and often disgusting, when the country goes into a brief economic decline and these people wring their hands, bellyache, and say it's the worst time they've ever seen. If they had the depression times of their parents and grandparents, they would put a torch to the cities every night.

A fellow told me he can remember when in his small town there wasn't a pickup vehicle to be seen. Now they're everywhere, some the four-wheel-drive kind.

During one of these economic recessions, the bellyachers were at it again. As I was driving past a tavern located at a wide place in the highway, there were thousands of dollars' worth of late model cars and snowmobiles parked outside, this in an area known to be not the richest in the country.

We're told poverty and lack of jobs are the reason for the failed marriages and the breakup of the family. Have they ever considered prosperity? It appears it takes more character to cope with prosperity than with poverty. Today's American middle-class majority can, with their credit cards, debt, and some pretty good incomes, live the life once only experienced by millionaires.

When some high school students living in a medium-sized city were asked what was the most important thing in life, 10 percent took money over family. Their main complaint was the high cost of car insurance. Build a high school today, and you'd better purchase enough land to park those student cars with their radios and tape players.

This along with all those school buses on the road. No more walking miles to school as their grandparents and great-grandparents did. Some even in grade school stagger onto those buses drunk on booze or high on drugs, and this the first thing in the morning.

Biotechnology will have a greater impact on American society than anything that went before it in history. It's going to make its share of millionaires. Biotech companies are the darlings of Wall Street. One company on the first day of trading in its stock went from 30 to 90.

I've concentrated on economics and its influence on future medical research to make a point. The bottom line is and always will be money. While self-preservation will play a role, economics still determines all.

It is said work, drugs, religion, and sex are used by people to relieve the pain of their existence. Add television to that list. But it's economics that fills the gaps.

Does society have the right to deny children healthier bodies when better ways to produce the human species are found, including an artificial womb replacing pregnancy? But will the dilemmas caused by these ways outweigh the benefits? Or are they the solution, temporary, if not permanent, as we continue to pull one foot after another out of this muddy path of evolution called progress?

Do we have the right to deny women freedom from fear of sex abuse and rape when the sex urge can be brought under control? Little do people realize how true it is when they say, "Sex is all in the head." Will manipulating it up there be what controls the sex urge? Is love without sex worth the tradeoff of having full equality for women and their no longer being thought of as sex objects? But are neuters what we want, or are they inevitable as we continue down that muddy path of evolution called progress?

Can the family unit as we know it of the past and present survive? If the breakup of the family comes about, will it be because the people lacked the self-discipline and responsibility to maintain the ideal of the family, or because they were unable to endure and cope with the stress and strain put on the family by the complexity of a high-tech modern age? Will the family's resiliency finally be broken as we continue down that muddy path of evolution called progress?

Volumes will be written about what I've just touched upon. My book will seem only minute in comparison. The question is, when do we as a society finally face up to the issue? Some say something cannot become part of society's dialogue until it actually becomes a reality. Then people adapt to whatever is demanded of them.

Energy became an issue and then a crisis when gas more than tripled in price and in some cases people waited in lines to fill their gas tanks. They adapted. Now if they get a gallon of gas for five cents cheaper they feel it's a deal and drive out of their way to buy it.

The energy crisis of the 1970s will seem like a picnic in comparison to what is coming in biotechnology. People will adapt and feel they're living better than ever. How can they not feel that way with all the advantages coming with living in the future? Still, it will be no utopia.

The media say you have to give the people what they want and not what they should know. That's what determines the media's profits and ratings. So they continue to give people gossip, trivia, and irrelevance. Maybe "Star Trek" is television's contribution to relevance. After all, who thought those comic strips of Flash Gordon and Buck Rogers years ago and their ventures into space would become a reality?

Probably I will not be popular for what I've written in this book. A comparison could be made to a woman told her husband is cheating on her, who reacts by saying that it's not true or she doesn't want to hear about it. The response to such news is usually anger. When the truth is finally accepted, more often than not, it still won't win the woman's friendship. Rational persons no doubt were burned at the stake as witches in the past because people at the time could not comprehend these individuals' foresight.

As I wrote in the beginning, this was written for those who will be living in the next century, the twenty-first century. As history has so often proved, one is not appreciated by his contemporaries if he is ahead of the crowd.

I end, as I also wrote in the beginning, by stating that history has shown reality often begins as an absurdity. So it shall be.